Energy Sector Standard of the People's Republic of China

NB/T 10231-2019

Basic specifications for the multiple paths ultrasonic flowmeter of hydropower stations

水电站多声道超声波流量计基本技术条件

(English Translation)

China Water & Power Press
中国水利水电出版社
Beijing 2024

图书在版编目（CIP）数据
水电站多声道超声波流量计基本技术条件 : NB/T 10231-2019 = Basic specifications for the multiple paths ultrasonic flowmeter of hydropower stations （NB/T 10231-2019） : 英文 / 国家能源局发布. 北京 : 中国水利水电出版社, 2024. 10. -- ISBN 978-7-5226-2759-5
Ⅰ. TV74-35
中国国家版本馆CIP数据核字第2024DD5441号

Energy Sector Standard of the People's Republic of China
中华人民共和国能源行业标准

Basic specifications for the multiple paths ultrasonic flowmeter of hydropower stations

水电站多声道超声波流量计基本技术条件

NB/T 10231-2019

(English Translation)

Issued by National Energy Administration of the People's Republic of China
国家能源局　发布
Translation organized by China Renewable Energy Engineering Institute
水电水利规划设计总院　组织翻译
Published by China Water & Power Press
中国水利水电出版社　出版发行
Tel: (+ 86 10) 68545888　68545874
sales@mwr.gov.cn
Account name: China Water & Power Press
Address: No.1, Yuyuantan Nanlu, Haidian District, Beijing 100038, China
http: //www.waterpub.com.cn
中国水利水电出版社微机排版中心　排版
北京中献拓方科技发展有限公司　印刷
210mm×297mm　16 开本　1.5 印张　60 千字
2024 年 10 月第 1 版　2024 年 10 月第 1 次印刷
Price（定价）: ¥ 240.00

About English Translation

This English version is one of China's energy sector standard series in English. Its translation was organized by China Renewable Energy Engineering Institute authorized by National Energy Administration of the People's Republic of China in compliance with relevant procedures and stipulations. This English version was issued by National Energy Administration of the People's Republic of China in Announcement [2023] No. 5, dated October 11, 2023.

This version was translated from the Chinese Standard NB/T 10231-2019, *Basic specifications for the multiple paths ultrasonic flowmeter of hydropower stations*, published by China Water & Power Press. The copyright is reserved by National Energy Administration of the People's Republic of China. In the event of any discrepancy in the implementation, the Chinese version shall prevail.

Many thanks go to the staff from the relevant standard development organizations and those who have provided generous assistance in the translation and review process.

For further improvement of the English version, any comments and suggestions are welcome and should be addressed to:

China Renewable Energy Engineering Institute
No. 2 Beixiaojie, Liupukang, Xicheng District, Beijing 100120, China
Website: www.creei.cn

Translating organizations:

POWERCHINA Northwest Engineering Corporation Limited

China Renewable Energy Engineering Institute

Translating staff:

YAN Cunku LI Kejia XIE Yonglan ZONG Wanbo

Review panel members:

YAN Wenjun	Army Academy of Armored Forces, PLA
QI Wen	POWERCHINA Beijing Engineering Corporation Limited
QIE Chunsheng	Senior English Translator
LI Nan	POWERCHINA Guiyang Engineering Corporation Limited
ZHAO Xin	Xi'an High Voltage Apparatus Research Institute Co., Ltd.
GUO Pengcheng	Xi'an University of Technology
WANG Yuchuan	Northwest A&F University
LI Shisheng	China Renewable Energy Engineering Institute

National Energy Administration of the People's Republic of China

翻译出版说明

本译本为国家能源局委托水电水利规划设计总院按照有关程序和规定，统一组织翻译的能源行业标准英文版系列译本之一。2023 年 10 月 11 日，国家能源局以 2023 年第 5 号公告予以公布。

本译本是根据中国水利水电出版社出版的《水电站多声道超声波流量计基本技术条件》NB/T 10231—2019 翻译的，著作权归国家能源局所有。在使用过程中，如出现异议，以中文版为准。

本译本在翻译和审核过程中，本标准编制单位及编制组有关成员给予了积极协助。

为不断提高本译本的质量，欢迎使用者提出意见和建议，并反馈给水电水利规划设计总院。

地址：北京市西城区六铺炕北小街 2 号
邮编：100120
网址：www.creei.cn

本译本翻译单位：中国电建集团西北勘测设计研究院有限公司
水电水利规划设计总院

本译本翻译人员：严存库　李可佳　谢永兰　宗万波

本译本审核人员：

闫文军　中国人民解放军陆军装甲兵学院

齐　文　中国电建集团北京勘测设计研究院有限公司

郄春生　英语高级翻译

李　南　中国电建集团贵阳勘测设计研究院有限公司

赵　鑫　西安高压电器研究院股份有限公司

郭鹏程　西安理工大学

王玉川　西北农林科技大学

李仕胜　水电水利规划设计总院

国家能源局

Contents

Contents

Foreword

This standard is drafted in accordance with the rules given in the GB/T 1.1-2009 *Directives for standardization—Part 1: Structure and drafting of standards.*

National Energy Administration of the People's Republic of China is in charge of the administration of this standard. China Renewable Energy Engineering Institute is responsible for its routine management. Energy Sector Standardization Technical Committee on Hydropower Hydraulic Machinery is responsible for the explanation of specific technical contents. Comments and suggestions in the implementation of this standard, if any, should be addressed to:

China Renewable Energy Engineering Institute
No. 2 Beixiaojie, Liupukang, Xicheng District, Beijing 100120, China

Drafting organizations:

POWERCHINA Northwest Engineering Corporation Limited

NARI Technology Co., Ltd.

Nanjing Sunrise Electric System Control Co., Ltd.

Dongfang Electric Machinery Co., Ltd.

Chief drafting staff:

DENG Zhiyong	XIE Yonglan	YUAN Lianjun	ZHANG Junzhi
SONG Jianying	WANG Shaofeng	MA Duo	LIU Dong
XIA Zhou	XU Chunrong	XU Weiwei	

Introduction

The flow measurement of hydroelectric generating unit is important for monitoring and evaluating the unit operating conditions, and ultrasonic flowmeters have been widely used for flow measurement.

In order to unify the design, type selection, installation and test requirements for multiple paths ultrasonic flowmeters and ensure the accuracy, reasonableness and economy of flow monitoring, this standard has been developed in accordance with the requirements of Document GNKJ [2015] No. 12 issued by National Energy Administration of the People's Republic of China, "Notice on Releasing the Development and Revision Plan of the Second Batch of Energy Sector Standards in 2014", and after extensive investigation and research, summarization of practical experience, consultation of relevant Chinese standards, and wide solicitation of opinions.

Basic specifications for the multiple paths ultrasonic flowmeter of hydropower stations

1 Scope

This standard specifies the requirements for the performance, system composition and installation of multiple paths ultrasonic flowmeters for hydropower stations, and the associated test, inspection, packaging, transportation, etc.

This standard is applicable to the measurement of flow through hydraulic turbines, pump-turbines and storage pumps, and through the pressure pipeline with a pipe diameter or equivalent pipe diameter of 800 mm and above.

2 Normative references

The following referenced documents are indispensable for the application of this document. For dated references, only the edition cited applies. For undated references, the latest edition of the referenced document (including any amendments) applies.

GB 50168, *Standard for construction and acceptance of cable line electric equipment installation engineering*

GB/T 150.3, *Pressure vessels—Part 3: Design*

GB/T 191, *Packaging—Pictorial marking for handling of goods*

GB/T 6587, *General specification for electronic measuring instruments*

GB/T 13384, *General specifications for packing of mechanical and electrical product*

GB/T 17626.2, *Electromagnetic compatibility—Testing and measurement techniques—Electrostatic discharge immunity test*

GB/T 17626.4, *Electromagnetic compatibility—Testing and measurement techniques—Electrical fast transient/burst immunity test*

GB/T 17626.5, *Electromagnetic compatibility—Testing and measurement techniques—Surge immunity test*

GB/T 17626.8, *Electromagnetic compatibility—Testing and measurement techniques—Power frequency magnetic field immunity test*

GB/Z 35717, *Discharge measurement for hydraulic turbines, storage pumps and pump-turbines—Ultrasonic transient-time method*

3 Terms and definitions

For the purposes of this document, the following terms and definitions apply.

3.1

ultrasonic flowmeter

installation for flow measurement by using the propagation characteristics of ultrasonic waves in fluid, which is composed of one or several pairs of ultrasonic transducers, transducer supports, radio frequency (RF) cables, cable conduits, cable sealing devices and host systems, etc. The ultrasonic flowmeters are classified, by the installation mode of transducers, into plug-in type, built-in type and clamp-on type

3.2

ultrasonic transducer

device that can produce ultrasonic output under the action of electrical signals and convert ultrasonic signals into electrical signals

[GB/Z 35717-2017, definition 3.2]

3.3

ultrasonic path

path travelled by an ultrasonic signal between a pair of ultrasonic transducers

3.4

ultrasonic path angle

included angle between ultrasonic path and conduit axis

[GB/Z 35717-2017, definition 3.8]

3.5

ultrasonic path length

actual distance that ultrasonic propagates in water medium between the paired transducers

[GB/Z 35717-2017, definition 3.7]

3.6

multiple paths ultrasonic flowmeter

device equipped with multiple pairs of ultrasonic transducers, is used for flow measurement through the ultrasonic propagation time method

3.7

insolubles content

suspended insoluble particles in water and the amount of bubbles characterized by volume ratio

3.8

accuracy class

rank or grade of an ultrasonic flowmeter to keep the measurement errors within the specified limits under standard laboratory conditions

3.9

measurement uncertainty

parameter, associated with the result of a measurement, that characterizes the dispersion of the values that could reasonably be attributed to the measurand, including the components caused by the influences of the system and the random

4 Technical requirements

4.1 Operating conditions

4.1.1 Ambient temperature

The operating ambient temperature for ultrasonic transducer should be -25 °C to 55 °C, and the operating ambient temperature for host system should be 5 °C to 40 °C.

4.1.2 Ambient humidity

The operating ambient relative humidity for host system should not be greater than 95 %.

4.2 Performance requirements

4.2.1 Accuracy class

The accuracy class of ultrasonic flowmeter should not be lower than Grade 0.5.

4.2.2 Service conditions and measurement uncertainty

The measuring section of the ultrasonic flowmeter shall be set at the pipe section with a certain straight length both upstream and downstream. Before measuring, the headrace shall be fully filled with water, the average flow velocity at the measuring section should be 0.3 m/s to 20 m/s, and the insolubles content in water should be less than 1 %. Under the condition that the length of the straight pipe section is not less than $10D$ (D is the equivalent inner diameter of the measured pipe) at the upstream side and is not less than $3D$ at the downstream side, the flow measurement uncertainty should not be greater than 1 %. When the length of straight pipe section is less than $10D$ at the upstream side and is less than $3D$ at the downstream side, refer to Annex A for the evaluation of measurement uncertainty.

4.2.3 Insulation

The insulation resistance of the tested part of ultrasonic flowmeter (AC 220 V terminal and signal input terminal) to the grounding point of the case should not be less than 10 MΩ.

4.2.4 Pressure tightness

For plug-in type and built-in type ultrasonic flowmeters, the protection degree of ultrasonic transducer should not be lower than IP68. The withstand pressure should be higher than the ambient pressure.

If the host system panel is placed at the turbine floor of the powerhouse or lower elevation, the degree of protection provided by the enclosure of the host system should not be lower than IP54. If it is placed in the station panel room or above the turbine floor of the powerhouse, the degree of protection provided by the enclosure of the host system should not be lower than IP41.

4.2.5 Electromagnetic compatibility

Electromagnetic compatibility (EMC) of ultrasonic flowmeter shall meet the following requirements:

- **a)** The severity level of electrostatic discharge immunity shall reach Level 3 and the flowmeter shall work normally during the test.
- **b)** The severity level of electrical fast transient/burst immunity shall reach Level 2, and flowmeter is allowed to malfunction during the test, but shall be able to recover automatically after test.
- **c)** The severity level of power frequency magnetic field immunity shall reach Level 3 and the flowmeter shall work normally during the test.
- **d)** The severity level of surge immunity shall reach Level 2 and flowmeter is allowed to malfunction during the test, but shall be able to recover automatically after test.

4.2.6 Continuous current test

The ultrasonic flowmeters shall undergo a continuous current test for no less than 72 h before delivery from the factory.

4.2.7 Reliability index

Under the specified working conditions, the ultrasonic flowmeter shall have the following reliability indexes:

a) The mean time between failures (MTBF) shall not be less than 20000 h.

b) The service life shall not be less than 10 years.

c) The availability shall not be less than 99 %.

4.3 Functional requirements

4.3.1 Data display and human-machine interaction

The ultrasonic flowmeters shall be able to display, as a minimum, the instantaneous flow, cumulative flow, flow velocity, date and time, and have the functions of zeroing and zero point calibration.

The host of the ultrasonic flowmeter shall be operable through keyboard or touch screen.

4.3.2 Data transmission

The ultrasonic flowmeters shall have the functions of local display and teletransmission. The unit's flow may be output in pulse, analog or digital mode. The interface mode shall meet the user's requirements, and the working conditions and fault condition signals of the ultrasonic flowmeters should be output in switching quantity.

4.3.3 Self-Diagnosis

The ultrasonic flowmeters shall have self-diagnostic function and be able to judge the current working conditions.

4.4 Appearance requirements

The surface of the new ultrasonic flowmeter shall be uniform in color, and the antirust coating shall be free from peeling-off and flaking, etc. The protective glass on the measuring host shall be transparent, the keyboard shall be agile and accurate. The marks on each socket shall be obvious, and the connectors shall be firm and reliable, and free of looseness caused by vibration. The display shall be neat and obvious, and the letter symbols and signs shall be legible and regular.

4.5 Main components

4.5.1 Host system

The host system shall meet the following requirements:

a) The host system shall be equipped with ultrasonic transmitting and receiving system (data acquisition) and microcomputer-controlled system. All I/O interfaces shall adopt plug-in structure, and shall be standardized, modularized and easy to expand and replace.

b) Each module of the host system shall have channel and module status indication.

c) The storage unit of the host system shall be provided with power-failure protection, which can store specified quantity of history data.

d) Each set of ultrasonic flowmeters may be provided with a host system using a separate panel (cabinet), or several sets of ultrasonic flowmeters may share one set of host system. The host system shall be provided with an LCD display and operation keypad.

e) The host system shall be provided with one set of power module, which supplies working power for ultrasonic transducers and secondary instruments.

f) The host system shall have various communication interfaces and signal cable interfaces to meet the users communication interface requirements.

4.5.2 Ultrasonic transducer

The 1 MHz ultrasonic transducer should be used when the ultrasonic path length is less than 10 m, while 500 kHz ultrasonic transducer should be used when the ultrasonic path length is more than 10 m. For the built-in type ultrasonic flowmeter, the ultrasonic transducer should adopt primary and backup element configuration. The plug-in type ultrasonic transducer shall allow replacement without emptying the pipeline.

4.5.3 Radio frequency (RF) cable

The ultrasonic transducer should be connected to the host system through coaxial RF cable which should not be extended through a connector.

4.6 Type selection of ultrasonic flowmeters

4.6.1 The plug-in type ultrasonic flowmeter is such an ultrasonic flowmeter that the ultrasonic transducer is inserted from the outside of the flow conduit through a drill hole and contacts with the measured medium. The built-in type ultrasonic flowmeter is such an ultrasonic flowmeter that the ultrasonic transducer is fixed on the inner wall of the pipeline and the signal cable is led out of the pipe through the cable penetrator and embedded cable conduit. The clamp-on type ultrasonic flowmeter is such an ultrasonic flowmeter that the ultrasonic transducer is fixed on the outer wall of the pipeline and does not directly contact with the measured medium. The clamp-on type ultrasonic flowmeter should not be used for long-term measurement of unit discharge.

4.6.2 When the measuring section is set at the exposed penstock section, the plug-in type ultrasonic flowmeters should be selected; when it is impossible, built-in type ultrasonic flowmeter should be selected.

4.6.3 As the installation of plug-in type ultrasonic flowmeter and built-in type ultrasonic flowmeter needs hole drilling on the penstock, the hole making and reinforcing shall be in accordance with GB/T 150.3.

4.7 Layout of ultrasonic paths

4.7.1 Selection of the number of ultrasonic paths

The selection of the number of ultrasonic paths of ultrasonic flowmeter shall meet the following requirements:

a) The number of ultrasonic paths for each ultrasonic path surface of ultrasonic flowmeters used for unit's flow measurement shall not be less than 4.

b) The primary basis for selecting the number of ultrasonic paths is the requirement of measurement uncertainty and the flow condition of the medium in the measured pipe section (depending on the upstream flow obstacles and straight-pipe section conditions) and the size of pipe diameter, refer to Annex A.

4.7.2 Layout pattern of ultrasonic paths

The ultrasonic path layout pattern of multipath ultrasonic flowmeters should be parallel or crossing/adopt parallel or crossing way. Figure 1 is the schematic diagram of crossing four ultrasonic paths (eight ultrasonic paths) and parallel four ultrasonic paths.

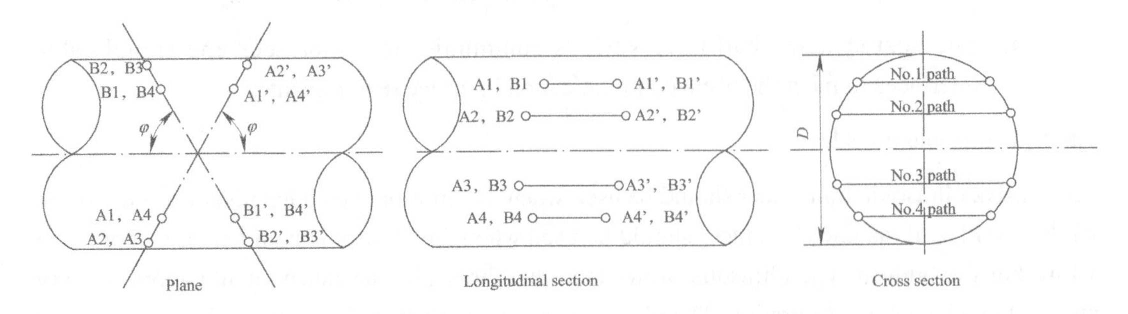

a) Layout of crossing four ultrasonic paths (eight ultrasonic paths)

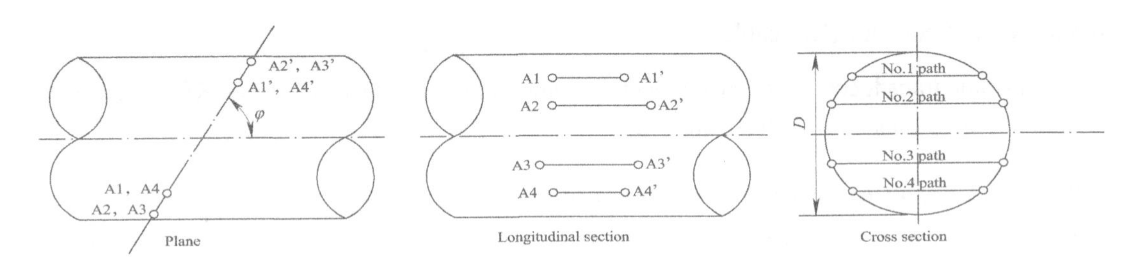

b) Layout of parallel four ultrasonic paths

Key

D	equivalent inner diameter of measured pipe
φ	ultrasonic path angle
A1 - A4 (A1' - A4'), B1 - B4 (B1' - B4')	measuring points

Figure 1 Schematic diagram of ultrasonic path layout

4.8 Installation requirements

4.8.1 Selection of installation position

The ultrasonic flowmeter should be set at the horizontal pipe section with steady water flow. If horizontal pipe section is not available, the inclined pipe section or vertical pipe section may also be selected should the installation requirements be satisfied.

4.8.2 Installation of ultrasonic transducer

The installation position of ultrasonic transducers should avoid the welding position, and the shape of ultrasonic transducer support shall be smooth to reduce the influence on the flow regime in the pipeline. The technical requirements for installation shall consider the site conditions.

4.8.3 Cable lead-out and installation of embedded cable conduit for built-in type ultrasonic transducer

Cable lead-out of built-in type ultrasonic transducer and installation mode of embedded cable conduit are shown in Figure 2.

The cable lead-out and the installation of embedded cable conduit for built-in type ultrasonic transducer shall meet the following requirements:

a) When installing the built-in type ultrasonic transducer, the distance from the circumferential cable conduit to the nearest ultrasonic transducer shall be greater than 1D (D is the diameter of the measured pipe), and when the ratio of the cable conduit diameter to the flow diameter is greater than 1 : 50, the distance shall be greater. For pump-turbines, the distance shall be greater than 2D.

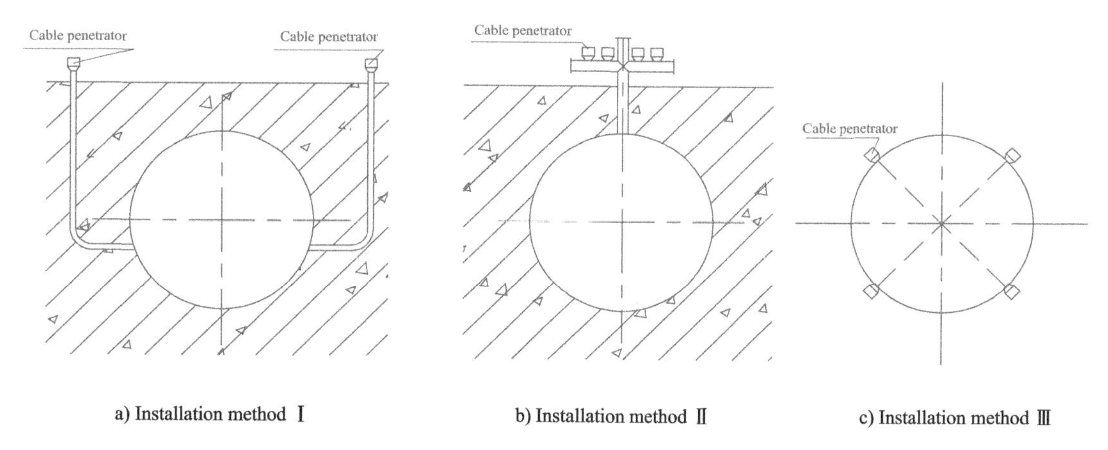

a) Installation method Ⅰ　　b) Installation method Ⅱ　　c) Installation method Ⅲ

NOTE 1 Installation methods Ⅰ and Ⅱ are both embedded cable conduits.

NOTE 2 Installation method Ⅲ is to provide cable penetrators in the exposed penstock section without embedding cable sleeves.

Figure 2 Schematic diagram of cable lead-out and embedded cable conduit installation of built-in type ultrasonic transducer

b) If the installation section of the ultrasonic transducers is in or close to the exposed penstock section, the cable penetrator may be directly installed on the penstock.

c) If the signal cable sleeve needs to be set on the penstock, the diameter of the sleeve depends on the number of ultrasonic transducers.

d) Guard plates shall be set at the RF cables sleeve to avoid direct scouring by water flow. The guard plates shall be specially designed and their appearance shall be smooth. Special sealing device shall be provided at the hole where the RF cable is led out, and no water leakage is allowed.

e) Other requirements for cable laying and installation of embedded cable conduits shall be in accordance with GB 50168.

5 Inspection method

5.1 Environmental conditions for inspection

Unless otherwise specified, the ultrasonic flowmeters should be inspected under the following conditions:

a) Ambient temperature: 5 °C to 40 °C for host system and -25 °C to 55 °C for ultrasonic transducer.

b) Relative humidity for host system: not greater than 95 %.

5.2 Visual inspection

The visual inspection of ultrasonic flowmeters shall meet the requirements given in 4.4.

5.3 Operating environment

The temperature and humidity test shall be conducted as per GB/T 6587.

5.4 Functional inspection

5.4.1 Preparations before functional inspection

Before the functional test the ultrasonic transducers shall be fixed in still water in pairs and be networked with the host, and be checked item by item in accordance with the requirements in 4.3.

5.4.2 Data display and human-machine interaction

When operating the ultrasonic flowmeter host by keyboard or touch screen, the display content and functions shall be complete and stable, the interface switching shall be smooth, and the key operation shall be agile.

5.4.3 Data transmission

Data transmission shall meet the following requirements:

a) If digital communication is used, the itemwise inspection shall be carried out in accordance with the communication protocol through running the communication software on a PC.

b) If pulse or analog communication is used, the inspection shall be carried out by process calibrator or oscilloscope.

5.4.4 Self-Diagnostic function check

The self-diagnostic function check shall meet the following requirements:

a) Check if the sound velocity in still water is consistent with the theoretical sound velocity at the current water temperature, and the deviation shall be less than 0.2 %.

b) Check if the indication of the working state truly reflects the current actual working state of the device.

5.5 Performance inspection

5.5.1 Accuracy inspection

When the ultrasonic flowmeters are delivered from the factory, the accuracy testing may be carried out by the hydrostatic sound velocity method under laboratory conditions. As specified by JJG 1030, the installed ultrasonic flowmeters may be tested online based on the comparison of sound velocities.

5.5.2 Insulation inspection

For ultrasonic flowmeters, use a 500 V megger to measure the insulation resistance of the tested part (AC 220 V terminal) to the grounding point of the case, which shall meet the requirements as specified in 4.2.3.

5.5.3 Pressure tightness

Place the probe of the ultrasonic transducer into the pressure tank, make sure that no water enters the tail of the cable, pressurize the tank to a pressure which shall not be less than 1.5 times the operating ambient pressure. The ultrasonic transducer or cable shall be able to withstand the pressure for 1 h without seepage, and the pressure drop of the pressure tank shall not exceed 0.1 MPa.

5.5.4 Electromagnetic compatibility

Electromagnetic compatibility shall meet the following requirements:

a) The electrostatic discharge immunity test shall be conducted in accordance with GB/T

17626.2, and the results shall meet the requirements as specified in 4.2.5.

b) The electrical fast transient/burst immunity test shall be conducted in accordance with GB/T 17626.4, and the results shall meet the requirements as specified in 4.2.5.

c) The power frequency magnetic field immunity test shall be conducted in accordance with GB/T 17626.8, and the results shall meet the requirements as specified in 4.2.5.

d) The surge immunity test shall be conducted in accordance with GB/T 17626.5, and the results shall meet the requirements as specified in 4.2.5.

5.5.5 Long-term stability inspection

Connect the ultrasonic flowmeter host and its associated ultrasonic transducers into a system. The system shall be subjected to a 72 h continuous current test in still water after the ultrasonic signal is adjusted in accordance with the instruction manual. The equipment shall work normally during inspection.

6 Inspection and test

The factory inspection and site test shall be carried out for each ultrasonic flowmeter, and the items should be in accordance with Table 1.

Table 1 Factory inspection and site test items

S/N	Item	Requirement	Method	Factory inspection	Site test
1	Appearance	See 4.4	See 5.2	√	√
2	Data display and human-machine interaction	See 4.3.1	See 5.4.2	√	√
3	Data transmission	See 4.3.2	See 5.4.3	√	√
4	Self-Diagnostic function	See 4.3.3	See 5.4.4	√	√
5	Accuracy	See 4.2.1	See 5.5.1	√	√
6	Insulation	See 4.2.3	See 5.5.2	√	–
7	Pressure tightness	See 4.2.4	See 5.5.3	√	–
8	Continuous current	See 4.2.6	See 5.5.5	√	√

7 Packaging, transportation and storage

Packaging, transportation and storage of ultrasonic flowmeter shall meet the relevant requirements of GB/T 191 and GB/T 13384.

Annex A
(informative)
Evaluation of measurement uncertainty of ultrasonic flowmeters

A.1 Reference value of uncertainty introduced by site flow conditions

As specified in JJF 1358, the reference value of uncertainty introduced by site flow conditions may be taken from Table A.1.

Table A.1 Reference value of uncertainty introduced by site flow conditions

Distance from flow obstacle L	Number of ultrasonic paths for single-measuring plane n	Number of measuring planes	
		Single-Plane	Double-Plane
$1D < L \leq 3D$	$3 < n \leq 5$	–	2 %
	$5 < n \leq 7$	2 %	1.5 %
	$n \geq 8$	1.5 %	1 %
$3D < L \leq 6D$	$3 < n \leq 5$	–	1.5 %
	$5 < n \leq 7$	2 %	1 %
	$n \geq 8$	1.2 %	0.7 %
$6D < L \leq 10D$	$3 < n \leq 5$	1.5 %	1 %
	$5 < n \leq 7$	1 %	0.8 %
	$n \geq 8$	0.7 %	0.5 %
$10D < L < 20D$	$3 < n \leq 5$	1 %	0.8 %
	$5 < n \leq 7$	0.8 %	0.5 %
	$n \geq 8$	0.5 %	0.3 %

NOTE 1 The above reference values are obtained through numerical simulation calculation of Gauss-Jacobi integral model of the flowmeter and conservative estimate, based on the fact that the flow obstacles are single elbows, located upstream of the flowmeters, with flow velocity ranging from 1 m/s to 7 m/s in smooth flow channels.

NOTE 2 If the specific flow conditions on site deviate from the above conditions, the given values can be used for reference.

A.2 Reference value of uncertainty introduced by transducer bulge

The ultrasonic transducer bulge will affect the uncertainty of ultrasonic flowmeters. The larger the bulge ratio is, the greater the measurement uncertainty. Ultrasonic transducer bulge influence should be evaluated or corrected. Please refer to GB/Z 35717 for the uncertainty evaluation of ultrasonic transducer bulge.

Bibliography

[1] JJG 1030, *Verification regulation of ultrasonic flowmeters*

[2] JJF 1358, *Calibration specification for DN1 000~DN15 000 liquid ultrasonic flowmeters calibration by non practical flow method*

微信号：Waterpub-Pro

微信号：悦读水电

唯一官方微信服务平台

销售分类：水电工程

ISBN 978-7-5226-2759-5

定价: 240.00 元

ICS 27.160
CCS F 12

Energy Sector Standard of the People's Republic of China

NB/T 10394-2020

Specification for photovoltaic power generation system performance

光伏发电系统效能规范

(English Translation)

Issue date: 202010-23 Implementation date: 2020-10-23

Issued by National Energy Administration of the People's Republic of China